Luana Soares da Silva
Bruno Cordeiro de Almeida
Raimundo Nonato Távora Costa

Biochemical attributes of papaya on ash and mulch

Luana Soares da Silva
Bruno Cordeiro de Almeida
Raimundo Nonato Távora Costa

Biochemical attributes of papaya on ash and mulch

Quality of the papaya tree

ScienciaScripts

Imprint

Cover image: www.ingimage.com

This book is a translation from the original published under ISBN 978-3-330-99608-3.

Publisher:
Sciencia Scripts
is a trademark of
Dodo Books Indian Ocean Ltd. and OmniScriptum S.R.L publishing group

120 High Road, East Finchley, London, N2 9ED, United Kingdom
Str. Armeneasca 28/1, office 1, Chisinau MD-2012, Republic of Moldova, Europe
Managing Directors: Ieva Konstantinova, Victoria Ursu
info@omniscriptum.com

Printed at: see last page
ISBN: 978-620-8-58846-5

SUMMARY

SUMMARY

The research was carried out in a family farmer's area in the municipality of Pentecoste, Ceará, from January 2014 to May 2015 and its aim was to assess the qualitative characteristics of total soluble solids, pH, total titratable acidity and the TSS(°Brix)/Acidity ratio of formosa papaya fruit in an experimental unit subjected to the production factors vegetable ash and mulch. An experimental unit was set up in a randomised block design with subdivided plots, consisting of four primary treatments (doses of ash), two secondary treatments (with and without mulch) and four blocks. The primary treatments corresponded to the following doses of vegetable ash: 50, 100 and 150% of the recommended amount of the production factor, as well as a control treatment (without application of the input). The secondary treatment consisted of the use of mulch based on carnauba bagana at a rate of 16 tonnes ha-1 and a control treatment (no mulch). Water was applied to the crop using a localised micro-sprinkler irrigation system with a non-deficit irrigation strategy. The results led to the following conclusions: The fruit of the Formosa Group papaya showed average values associated with the qualitative variables total soluble solids (TSS), total titratable acidity (TTA), pH and TSS/ATT ratio, which give them qualities within the standards for commercialisation on the domestic market. In addition, non-formal information obtained from consumers of this product is very positive about the quality of the fruit, which has helped the farmer to sell his product at an attractive price (R$2.00 kg-1). The degree

The sweetness of papaya fruit measured by the TSS/ATT ratio gives average values, if all the treatments (primary and secondary) are taken into account, that are up to double the average reference value for Formosa Group fruit. The non-significance of the primary and secondary treatments related to the qualitative variables of the papaya fruit analysed in this research is highly significant from the point of view of the allocation of financial resources in the factor-product relationship. The analyses related to the application of the plant ash production factor cannot be considered absolutely decisive due to problems associated with the application of this input during the experiment. A balance of nutrients required and made available, particularly nitrogen, points to the need to adjust the production system in order to maximise other qualitative variables of interest.

CHAPTER 1

INTRODUCTION

Family farming demonstrates its weight in the Brazilian food basket, given that it is responsible for 87% of national cassava production, 70% of bean production, 46% of corn, 38% of coffee, 34% of rice, 21% of wheat and, in livestock farming, 58% of milk, 59% of pigs, 50% of poultry and 30% of cattle. Despite cultivating a smaller area with crops and pastures (17.7 and 36.4 million hectares, respectively), family farming is responsible for guaranteeing a large part of the country's food security, as an important supplier of food for the domestic market (IBGE, 2006).

In the Curu Pentecoste Irrigated Perimeter, coconut and banana trees are grown mainly in a conventional production system, with farmers having no technical assistance and using inputs in large quantities, mainly with the indiscriminate use of pesticides and without any protection, and disposing of empty containers in the soil and irrigation channels.

In this context, the need for an agroecological transition has been highlighted, with the aim of achieving a more natural and sustainable production based on ecological principles. As a strategy to materialise the agroecological transition process, an experimentation unit can be used so that farmers can see the techniques used and observe the economic, social and environmental results, and thus evaluate the advantages and disadvantages of the system and compare it with what is traditionally used.

In the semi-arid region, which is characterised by water scarcity and a high evaporation rate, it is necessary to use more sustainable practices that save more water for irrigation. Mulching is a viable practice because, as well as reducing soil temperature and evaporation, it improves the macro and micro fauna in the plant's root zone as the material decomposes, providing mineralisation of some nutrients and thus bringing benefits to the plants.

Ash is used as a source of nutrients for plants and is a way of recycling the waste left over from burning wood in bakeries and steakhouses. This practice can replace chemical fertilisers that often leach into the soil. The use of this alternative input has high levels of K, P, Ca and Mg in its composition, allowing a nutritional balance to be made (NKANA et al., 1998).

The papaya tree (Carica papaya L.) is very well adapted to the tropical region, playing a major role in the domestic and foreign markets and is of great importance to family farmers where there are already successful experiences of growing it in the Curu Valley.

The aim of this research was to assess the qualitative characteristics of total soluble solids, pH, total titratable acidity and the TSS(°Brix)/Acidity ratio of formosa papaya fruit in a family-based experimentation unit subjected to the factors of production, vegetable ash and mulch.

CHAPTER 2

LITERATURE REVIEW

2.1 Family farming

According to Paduan (2014), family farming can be defined as agricultural production that is carried out by family members and that this production is for their own survival, generating a surplus that can be commercialised.

According to Law No. 11.326 of July 2006, family farming is defined as:

> Art. 3 For the purposes of this Law, a family farmer and a rural family entrepreneur are considered to be those who carry out activities in rural areas while simultaneously meeting the following requirements:
>
> I - does not own, in any capacity, an area larger than 4 (four) fiscal modules;
>
> II - use predominantly their own family's labour force in the economic activities of their establishment or enterprise;
>
> III - have a family income predominantly derived from economic activities linked to the establishment or enterprise itself;
>
> IV - run your establishment or business with your family.
>
> § Paragraph 1 - The provisions of item I of the caput of this article shall not apply in the case of rural condominiums or other collective forms of ownership,

provided that the ideal fraction per owner does not exceed 4 (four) fiscal modules.
§ Paragraph 2:

I - foresters who simultaneously fulfil all the requirements referred to in the heading of this article, cultivate native or exotic forests and promote the sustainable management of those environments;

II - aquaculture producers who simultaneously meet all the requirements referred to in the heading of this article and exploit water reservoirs with a total surface area of up to 2 ha (two hectares) or occupy up to $500m^3$ (five hundred cubic metres) of water, when the exploitation is carried out in net-tanks;

III - extractivists who simultaneously fulfil the requirements set out in items II, III and IV of the caput of this article and carry out this activity by hand in rural areas, excluding prospectors and miners;

IV - fishermen who simultaneously fulfil the requirements set out in items I, II, III and IV of the caput of this article and carry out fishing activities by hand.

According to Buainain (2004), Brazilian family farmers have differed throughout history due to the cultural legacies of each region, which are quite varied, the experiences they have had and the access they have had to factors such as natural resources, individual and collective development. These differences lie in these farmers' access to the market, which will change according to the development of the economy and the public policies of each region.

Also according to Buainain (2004), family farming includes poorer farmers, with a greater concentration in the municipalities of the

Northeast Region and the North of Minas Gerais. Although the Northeast is made up of a greater number of poor family farmers, they are very important in the agricultural production of the municipalities.

Family farming must be integrated with agro-industry chains, especially in niche markets or alternative commercialisation channels, such as organic production, which is increasingly expanding and gaining market share (BUAINAIN, 2006).

2.2 Alternative Farming and Agroecological Transition.

According to Ehlers (1996), after the two great world wars there was a great deal of technological development and some substances that were used in the wars were adapted for plant diseases and pests. After the Second World War, many companies adapted to manufacture agrochemicals and agricultural machinery. At the end of the 1960s and the beginning of the 1970s, great progress was being made in the agricultural industries and in research in the areas of chemistry, mechanics and genetic improvement, which was marked by transformations in history called the Green Revolution.

The Green Revolution was based on replacing local and traditional production practices with a set of technological practices using genetically modified varieties, high demand for synthetic chemical fertilisers, the use of pesticides, irrigation and mechanisation. During the Green Revolution, between 1950 and 1984, the availability of food increased by around 40%

per inhabitant, but despite this high agricultural production, there was great concern about the social, environmental and economic impacts (EHLERS, 1996).

In the second half of the 1970s, a group was formed in Brazil that challenged conventional agriculture, called Alternative Agriculture, made up of some agronomists from the city of São Paulo, characterised by a more critical outlook and concern for environmental problems. In the early 1980s, three Brazilian Alternative Agriculture Meetings were held, making them increasingly popular and increasing society's interest in environmental issues (KAMIYAMA, 2012).

From the 1980s onwards, various types of agriculture began, such as organic or biological, biodynamic, natural and permaculture, thus giving rise to agroecology, which is not only concerned with environmental issues, but also presents a broader vision and a new approach to agriculture (KAMIYAMA, 2012).

Organic farming is one of the most widespread alternative farming practices in the world. It became known after the work and observations of an agronomist, Sir Albert Howard, who wrote the book "An **Agricultural Testament", which** recounts all the successful experiences he had with peasants in India, highlighting the entire ecological basis and the importance of soil conservation and fertility, microorganisms that live in the soil, the role of organic matter in the soil, among others (PAULUS, 1999).

Biodynamics is an agriculture that was started by Ruldof Steiner at the beginning of the 20th century based on the principles of anthroposophy, which relates the universe to plants and man. It takes a holistic view of the world, relating cosmic influences to plants and animals, and also studies the interaction of these subtle energies with plants, animals and man (PAULUS, 1999).

Biodynamic agriculture uses astronomical knowledge to manage plants through biodynamic preparations and the Agricultural Astronomy calendar and with greater diversity in the agro-ecosystem, thus making the system more diverse and more sustainable on the property (KAMIYAMA, 2012).

Natural agriculture is one of the most radical in relation to conventional agriculture, having originated in Japan around 1935 by Mokiti Okada, with the aim of producing crops by copying nature. It is an agriculture based on the conservation and quality of the soil for a microbiological balance of the soil. Its agricultural practices are based on little soil movement such as green cover, no weeding and only mowing, intercropping with legumes, cocktails with vegetables and aromatic plants. It differs from the others in that it does not use manure, and its source of compost is only of vegetable origin (KAMIYAMA, 2012).

Permaculture emerged in the mid-1970s from the ideas of Bill Mollison and David Holmgren in Australia. This type of agriculture is concerned with making very efficient use of resources and sources of energy within the system, including animals in order to reuse their waste and to have greater diversity, not restricting the proposed changes mainly to the physical environment (PAULUS, 1999).

Agroecology emerged in the 1980s with the aim of studying all this alternative knowledge and bringing theoretical and methodological knowledge to the field of science to study the experiences of farmers. In Brazil, this term was adopted mainly by NGOs that had contacts with organic and family farmers (PAULUS, 1999).

Agroecology is considered a scientific discipline, but it is not a classical science that seeks accuracy. It recognises that knowledge can be generated in other contexts, such as traditional and cultural knowledge, which calls into question the references of classical science and the researchers themselves, not an easy task considering the training of researchers and the structures of institutions. (AQUINO; ASSIS,2005)

For the agroecological transition process to take place, the individuals who work in this agroecosystem need to consciously change their values and attitudes towards the management and preservation of natural resources. It is necessary to be aware that this transition process is complex, both from a technological, methodological and organisational point of view (CAPORAL; COSTABEBER, 2004).

According to Gliessman (2000), the agroecological transition is divided into three levels for converting to a sustainable agroecosystem: the first level is to reduce the use of external inputs in order to save on production costs and reduce environmental degradation; the second level would be to replace external inputs and conventional practices with alternative practices; and the third would be to search for the causes of problems and their solutions.

2.3 Importance, cultivation and characteristics of the papaya crop.

The papaya plant originated in the regions of southern Mexico and northern Nicaragua and was discovered by the Spanish, where it was distributed throughout much of tropical America and introduced to Brazil around 1587 (SERRANO; CATTANEO, 2010).

The papaya tree belongs to the class Dicotyledoneae, subclass Archichlamydeae, order Violales, suborder Caricaceae, family Caricaceae and genus Carica. It is a herbaceous plant with a green stem that can reach up to eight metres in height, with a commercial lifespan of up to four years (SIMÃO, 1998).

The Carica papaya species is the only one of commercial value, characterised by its unilocular ovary, pivoting root system with a well-developed main root and large branches in the first 0.30 m of the soil. It has alternate leaves, which grow in inflorescences that appear in the leaf axils (RIBEIRO, 2007).

The flowers of the papaya tree come in three types: female, male and hermaphrodite. Depending on the type of flower, it influences the shape of the fruit, with female flowers giving rise to larger, rounded fruits, while hermaphrodite flowers produce perforated fruits that have a higher commercial value (RIBEIRO, 2007).

The fruit is a berry that varies according to the type of flower, with a thin, smooth skin that protects a 2.5 to 5 cm pulp, the colour of which can vary from yellow to reddish (DANTAS; CASTRO, 2000).

According to the FAO (2010), world production of papaya is 10%

of tropical fruit, which means 8 million tonnes, with Brazil, Mexico, India and Indonesia being the main producers.

In 2013, Brazil produced 1.5 million tonnes of papaya, with the Northeast region being the largest producer with 978,000 tonnes (IBGE, 2015).

2.4 Nutrition and mulching.

Around the 19th century, agriculture was driven by technological changes in which Justus Von Liebig (1803 - 1873) postulated the **"Law of the Minimum".** He observed that plants thrived without the use of organic matter and that crop yields were a direct and proportional function of the quantities of mineral nutrients made available to plants, which could be calculated. Some even thought that organic fertilisers were useless for plants, i.e. that chemical fertilisation with synthetic fertilisers was intended to put an end to all the problems of soil depletion in agriculture, and that the soil was therefore just a substrate to support plants (ASSIS, 2002).

The use of alternative sources of fertiliser is important for reducing costs, increasing sustainability and preserving natural resources, and one of the main advantages of this fertilisation is the improvement of the physical and chemical aspects of the soil (PIVA *et.al,* 2014).

The papaya is a plant with very high nutritional requirements and because its roots are very shallow, they are more prone to nutritional deficiencies (TRINDADE, 2000).

Nitrogen (N) is the second most demanded nutrient by the papaya crop and is used by the plant mainly in its vegetative growth. It cannot be lacking in the six months after planting. Phosphorus (P) is required to a lesser extent than N and K, but its function is very important for the growth of the root system and also for fixing the fruit on the plant (TRINDADE, 2000).

Potassium (K) is the nutrient most required by papaya, and is needed increasingly throughout the plant's cycle. One of its most important functions is to provide larger fruit and high levels of sugar and total soluble solids after flowering. Calcium (Ca) is the third most required nutrient by the papaya tree, and its deficiency can lead to fruit with softened flesh and thus less resistance to transport and consequently less shelf life for marketing (OLIVEIRA, 2004).

Boron is the most important micronutrient for the papaya crop, and its deficiency can cause poor fruit formation (TRINDADE, 2000).

Alternative sources of fertiliser can be a way of replacing synthetic chemical fertilisers, a high percentage of which may not be absorbed by the plants and therefore contaminate groundwater, rivers and reservoirs.

Ash is an alternative source of nutrients, usually from burning wood, and can contain high levels of K, P, Ca and Mg (NKANA et *al,* 1998, *apud*, SOFIATTI, 2007).

Research results have shown that an increase in the amount of ash led to an increase in the soil's CEC, as well as in the levels of Ca^{2+}, Mg^{2+} and K^{+}, a fact that is due to the increase in the soil's pH, making the nutrients more available in the ash (CAMPANHARO et. *al,* 2008).

Mulching is a technique for protecting the soil from various interferences, such as solar radiation, rain, erosion and the high temperatures that occur mainly in the Northeast. Grasses are a way of mulching, as they form aggregates in the soil, increase aeration and favour the root system. Leguminous plants used as ground cover improve soil structure, increase biological nitrogen fixation and have little competition with fruit trees because their roots are shallow. Bagana, which comes from carnauba straw, is currently used in various types of fruit trees, in the projection of the canopy, as a way of reducing the evaporation of water in the soil, thus increasing the availability of nutrients and maintaining humidity in the plant's rhizosphere. It also reduces the occurrence of spontaneous plants that compete for nutrients with cultivated plants (OLIVEIRA, 2002).

Table 1 - Chemical and bromatological composition of carnauba bagana

Components	**%**
Dry Matter	86,00
Organic Matter	94,30
Ashes*	5,74
Crude protein*	8,04
Neutral detergent insoluble nitrogen/NT**	58,90
Acid detergent insoluble nitrogen/NT**	51,00
Neutral detergent fibre*	69,70
Acid detergent fibre*	51,20
Haemicellulose*	18,50
Cellulose*	39,90
Lignins*	10,60
Ether extract*	2,09
In vitro dry matter digestibility*	15,30
In vitro digestibility of organic matter*	11,70
Tannin*	12,20
Calcium*	0,16
Phosphorus*	0,13

Source: Gomes (2008).

NT - Total nitrogen; *Expressed on a dry matter basis; **Expressed on a total nitrogen basis.

2.5 Chemical attributes of papaya fruit.

During fruit development, physical and chemical changes occur which are important characteristics for the ripening process. The chemical composition is directly related to the stage of ripeness; the higher the total soluble solids and the lower the titratable acidity, the higher the degree of ripeness of the fruit (CARVALHO et al., 1992). Standards are established for the point of harvest and the quality of the fruit.

From a commercial point of view, the right time to harvest is generally determined by the distance from the consumer market, i.e. depending on how many days it takes to reach the market, as shown in Table 2.

Table 2 - Description of the main visual characteristics at the different stages of ripeness of *papaya* fruit *(Carica papaya* L.)

Harvest point	Visual characteristics
Stadium 0	• Grown and developed fruit (100% green); • The skin of the fruit turns from green to light green;
	- Yellowish white inner flesh with small light pink spots.
Stadium 1	• Known as the one-spot stage (15 per cent of the bark surface is yellow); • The skin of the fruit is lighter green; • An almost imperceptible yellow stripe

	(base of the fruit); • Stage used for fruit destined for export markets (by air) or for long distances (1000 to 2000 kilometres).
Stadium 2	• Two-spot stage (25% of the fruit surface is yellow); • Light green skin on the entire surface of the fruit; • Two very noticeable yellowish streaks (at the base); • Inner flesh coloured yellow in the area near the rind and reddish yellow near the ovaries; • Stage used for fruit exported by air or to domestic markets between 500 and 1000 kilometres from the place of production.
Stadium 3	• Three-spotted stage (50 per cent of the fruit surface is yellow); • 3 to 4 clearly visible yellowish streaks; • Reddish yellow flesh near the rind and orange-red near the ovaries; • Fruit sold to markets up to 500 kilometres from the place of production.

Source: Sanches (2003).

Total soluble solids (°Brix) are used to estimate the quantities of substances available in each fruit and are mainly composed of sugars (SHINAGAWA, 2009).

Some studies have shown that the standard SS content for papaya is between 11% and 14% during the winter and can reach 17% during the summer (ALVES *et. al,* 2003).

Oliveira, (1999) reports that Total Soluble Solids were 11.5% in the cultivar 'Tmproved Sunrise Solo Line 72/12' and 10.4% in the hybrid 'Tainung 01/781'.

Studies have shown that there was a significant difference in the amounts of SS in the different stages of ripeness, with the Tainung 01' hybrid having the highest levels in the second stage, which would be 25% of the surface of the yellow fruit (SOUZA, 2004).

Results obtained by Júnior et.al, (2007) show that for the Formosa Group, the Total Soluble Solids (TSS) variable is 14°Brix, the Titratable Acidity (TA) is 0.11, the pH is 5.20 and the ratio (TSS/TA) is 100.1 are average values.

Selvaraj et al. (1992) working with four types of papaya showed that as the fruit develops on the plant, there is a gradual increase in the amount of sucrose, fructose and glucose and that at the end of its growth the presence of fructose and glucose is predominant. (SHINAGAWA, 2009).

Total soluble solids (TSS) and titratable acidity are measures that indicate the point of harvest because there is a relationship between these parameters and the stage of ripeness of the fruit (FAGUNDES; YAMANISHI, 2001).

pH is a measure that represents the inverse of the amount of hydrogen ions (H^+) in a material. Weak free organic acids and potassium salts show a small variation in pH as a result of the stabilisation of the system. There is an increase in pH proportional to the reduction in acidity, and it is used as an indicator of this variation (CHITARRA; CHITARRA, 2005).

Acidity decreases as the fruit ripens and can be used with sweetness as a reference point for the degree of ripeness (CHITARRA; CHITARRA, 2005).

Soluble solids is a measure that is mainly used to quantify sugars and is one of the main chemical substances present in vegetables. The TSS content is measured using a refractometer, used mainly in the fruit processing industry as a control for correcting the sugar content of the product and the quantities of crystallised sugars that will be added depending on what is contained in the fruit (CHITARRA; CHITARRA, 2005).

CHAPTER 3

MATERIAL AND METHODS

3.1 Experiment characterisation

The experiment was conducted downstream of the Pereira de Miranda reservoir, in an area belonging to the National Department for Works Against Droughts - Dnocs, in the municipality of Pentecoste, Ceará, from January 2014 to February 2015.

The experiment area is located at the following UTM coordinates: (N; E) = (9579564; 0470725), zone 24 M and altitude of 25.0 metres. The region's climate is semi-arid and dry, with a short wet season that allows the development of low vegetation (Embrapa, 1999).

This research was carried out in an experimentation unit (Figure 1), which allows the farmer to observe the results obtained in practice and reflect on the advantages and disadvantages of the production system. It also allowed the exchange of knowledge with students and researchers from the Gpeas Group on a weekly basis, as a way of trying to improve social, environmental and economic aspects of the farmer's life.

Figure 1: Students from the Gpeas Group with the farmer in the experimentation unit.
Source: Author (2014).

The physical and chemical attributes of the soil in the 0 - 0.2 m layer (Table 3) were determined at the Soil and Water Laboratory of the Soil Science Department, part of the Agricultural Science Centre of the Federal University of Ceará.

Table 3 - Physical and chemical attributes of the soil in the arable layer (0 - 0.2 m)

Chemical attributes	(0 - 0,20 m)	Physical attributes	(0 - 0,20 m)
P2O5 (mg kg^{-1})	37,0	Fine sand (g kg^{-1})	706
K^{+} (cmolc kg^{-1})	0,32	Coarse sand (g kg^{-1})	40
Na^{+} (cmolc kg^{-1})	0,13	Silt (g kg^{-1})	175
Ca^{2+} (cmolc kg^{-1})	2,5	Clay (g kg^{-1})	39
Mg^{2+} (cmolc kg^{-1})	2,5	Natural clay (g kg^{-1})	39
Al^{3+} (cmolc kg^{-1})	0,1	Soil density (g cm^{-3})	1,62
M.O. (g kg^{-1})	7,97	Particle density (gcm^{-3})	2,64
C/N	10,0	Flocculation (g $100g^{-1}$)	51

pH	6,0	Textural class	Sandy loam
EC (dS m^{-1})	0,2	Humidity at 0.033 MPa	9,57
PST	2,0	Humidity at 1.5 MPa	4,74

Source: Soil and Water Laboratory of the Federal University of Ceará (2014).

Organic matter is of great importance in tropical and subtropical soils, as it influences their physical, chemical, biological and mineralogical attributes (DALMOLIN et al., 2005). In natural ecosystems, the organic matter content of the A horizon can reach 15 to 20 per cent or more. However, according to Gliessman (2001) what is found in most soils is an average of 1 to 5%, and the soil in the area of this experiment has an organic matter content that is slightly below this lower reference limit (0.8%).

The pH of the soil is within the recommended range for papaya (5.5 to 6.7), which is not very tolerant of acidity. However, too high a pH can cause reductions in productivity by reducing the availability of micronutrients such as boron, copper, iron, manganese and zinc (TRINDADE, 2000).

It is a sandy loam soil with no salinity or sodicity problems, and in the range of tensions between 33 and 1500 kPa it stores 15.6 mm in the first 0.20 m of soil, which exceeds the daily requirements corresponding to the irrigation rates in the most critical months of water demand.

With regard to the amount of fertiliser needed for the crop, the adjustment was made based on potassium and for the other nutrients, amounts outside the range indicated by Oliveira (2004) were used. A total of 587 g of Ca^{2+} per plant was used, far more than necessary, along with 55 g of Mg^{2+}. As for the nitrogen and phosphorus content, they were 129.87

and 50 g per plant, respectively, so they were in deficit in terms of the suggested proportion.

3.2 Experimental design and production factors

The experimental design was a randomised block with subdivided plots, consisting of four primary treatments, two secondary treatments and four blocks, as shown in Figure 2.

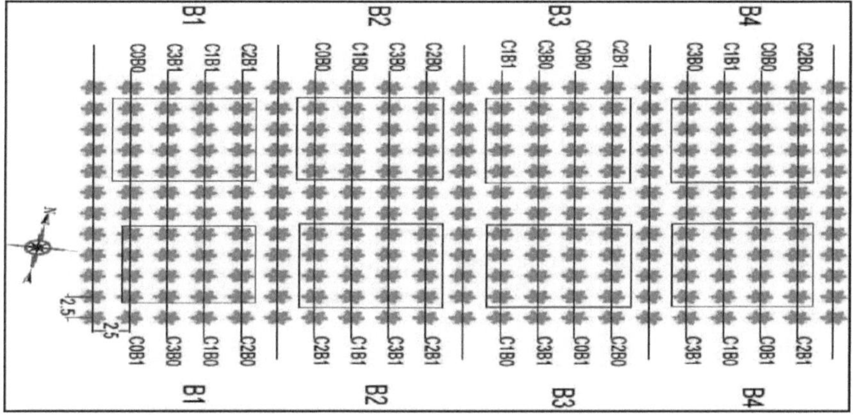

Figure 2 - Sketch of the experiment
Source: Author (2015).

The primary treatments consisted of levels corresponding to 0, 50, 100 and 150% of the recommended amount of vegetable ash, as well as a control treatment (no application of vegetable ash - C0), where the vegetable ash was produced in a bakery oven. Carnauba bagana was used as a secondary factor, with one treatment consisting of the application of 16 t ha^{-1} (B1) and a control treatment (no application of carnauba bagana - B0).The dose of ash corresponding to 100% of the requirement was 80 g of K_2O per plant, representing a total of 1195 g of vegetable ash in the first production cycle. The application was divided into five plots, as shown in Table 4, which shows the period of application and the respective quantities of vegetable ash recommended per application.

Table 4 - Period and amount of vegetable ash per plant for the respective treatments.

Period after planting (days) planned/executed	Dose of vegetable ash per plant (g)			
	0%	50%	100%	150%
30/40	0	112	224	336
90/125	0	112	224	336
150 /190	0	125	249	374
210/233	0	125	249	374
270/not executed	0	125	249	374

Source: Author, 2015.

Problems associated with logistics and availability meant that fertilisation was not carried out in accordance with the planned recommendations.

3.3 Establishing and managing the crop in the field

Before the papaya crop was established in the field, the legumes Calopogonium mucunoides and Pueraria phaseoloides were sown with a view to their incorporation and therefore greater availability of nutrients, especially nitrogen.

The Formosa papaya cultivar (Tainung 01) was sown in 168-cell black polyethylene trays on 8 June 2014, using a mixture of clay, sand and tanned cattle manure in proportions 1:1:1. Germination took place between 09 and 12 days after sowing.

The pits at a spacing of 2.5 m x 2.5 m were prepared manually to a size of 0.40 m x 0.40 m x 0.40 m and filled with 32 litres of tanned cattle and sheep manure and the same amount of topsoil taken from the pits themselves.

The chemical characteristics of the material used in the pit in a ratio of 50% tanned manure and 50% topsoil are shown in Table 5.

Table 5 - Chemical characteristics of the material used in the foundation fertiliser.

($g\ kg^{-1}$)							(mg kg-1)			
N	P	P_2O_5	K	K_2O	Ca	Mg	Fe	Ass	Zn	Mn
3,7	1,4	3,2	3,6	4,4	6,9	6	-	14,1	56,4	208,6

Source: Soil and Water Laboratory of the Federal University of Ceará (2014).

The final planting in the holes took place on 3 July 2014, 25 days after sowing, when the seedlings had six definitive leaves. During this period and for two days, the plants were exposed to full sun for adaptation purposes. Replanting took place six days after planting.

Figure 3: Start of planting (A) and plant development (B)

Source: Author (2015).

Figure 4: The papaya orchard.
Source: Author (2015).

Figure 5: Papaya orchard in production.
Source: Author (2015).

Cultivation treatments were carried out weekly, such as removing invasive plants by hand in the crown projection and with a hoe between plants and rows. Forty days after planting, the first fertiliser was applied with vegetable ash and mulched with carnauba bagana. 85 days after the first application, the second fertiliser with vegetable ash was applied. The third application of vegetable ash took place on 13 January 2015, 65 days after the second application. The fourth application took place on 21 February 2015, 233 days after planting and 39 days after the third application. However, the fifth application was not carried out, so the total effect of the recommended ash has not yet been seen.

3.4 Irrigation

A micro-sprinkler type localised irrigation system was used with one emitter for two plants operating at a flow rate of 112 L h^{-1} to at a working pressure of 15 mca, whose main line was made up of PVC pipes (nominal diameter 50 mm), with lateral lines made up of polyethylene hoses (nominal diameter 16 mm).

The water source comes from a small stream replenished by underground flow from the Pereira de Miranda reservoir, located near the experiment area, which is classified as C3S1 (Table 6), according to Ayers and Westcot (1999). It is therefore a quality of water that should not be used on soils with poor drainage and even with adequate drainage it should be used on salt-tolerant soils, using localised irrigation and appropriate management criteria.

Table 6 Chemical characteristics of the irrigation water

Cations (mmolc.L-1)					Anions (mmolc.L-1)					EC			SDT	
Ca^{2+}	Mg^{2+}	In^{+}	Kz^{+}		Cl^{-}	SO_4^{2-}	HCO_3^{-}	CO_3^{2-}		(dS m^{-1})	RAS	pH	(mg L^{-1})	Clas.
2,5	3,1	2,7	0,2	8,5	6,2	-	2,4	-	8,6	0,86	1,62	6,89	60	C3S1

Source: Soil and Water Laboratory of the Federal University of Ceará (2014).

The crop's water requirement was calculated for the different months and crop stages according to Equation 1:

$$Ti = \frac{ETo.Kc.Ap}{N.qe}$$

Ti: irrigation time, h; ETo: daily reference evapotranspiration, mm; Kc: crop cultivation coefficient; Ap: useful area per plant, m^2; qe: emitter flow rate, Lh^{-1}; N: number of emitters per plant.

Reference evapotranspiration (ETo) was calculated using the Penman-Monteih /FAO (1991) method, using the Cropwat software. The input data for calculating ETo was obtained from a historical series for the municipality of Pentecoste for the period 1970-1998, as shown in Table 7.

Table 7 - Average daily reference evapotranspiration for Pentecoste - CE

Month	Average daily ET0 (mm)
January	6,15
February	5,33
March	4,14
April	4,14
May	4,28
June	4,61
July	5,21
August	6,85
September	7,83
October	7,97
November	7,77
December	7,27

Source: Cabral (2000).

The crop coefficients for the various phenological stages were obtained by Montenegro, Bezerra and Lima (2004), as shown in Table 8.

Table 8 - Crop coefficients for papaya at the different phenological stages

Days after sowing	Stadiums	Kc
84-107	Vegetative	0,64
115-260	Flowering/Fruiting	1,16
267 - 380	Flowering/Fruiting/Maturation	1,19

Source: Montenegro, Bezerra and Lima (2004).

In the eighth month after planting, the fruit began to be harvested, and in the tenth month and at the second stage of ripeness (Figure 4), they were harvested for analysis of the variables pH, total soluble solids and titratable acidity at the Tropical Fruit Laboratory of the Federal University of Ceará.

Figures 6: Harvesting and weighing papaya.

Source: Author (2015).

Figure 7: Sorting and storing the papaya for transport.
Source: Author (2015).

3.5 Chemical attributes of papaya fruit

The fruit was sampled using one fruit from each treatment, making a total of 32 fruits analysed. The fruits were picked at the second stage of ripeness and taken to the laboratory.

Figure 8: Papaya samples for analyses.

Source: Author (2015).

3.5.1 pH

The research used an electrometric process in which specially adapted potentiometers are used to determine pH directly, simply and precisely. The following materials were used: 50 and 150 mL beakers, 200 mL beaker, peagameter, analytical balance, spatula, blender, distilled water, pH 4 and 7 buffer solutions.

In the procedure, 20g of the sample was weighed into a beaker and diluted with 200 mL of distilled water (1:10 dilution). The contents were stirred with using a blender until the particles were uniformly suspended. Three replicates were then carried out to determine the pH, using a previously calibrated pH meter.

3.5.2 Total Soluble Solids (TSS)

To determine soluble solids, a sample of pulp was taken from each repetition and processed in a Philips-Walita model RI 6720 mixer. The resulting juice was evaluated in a Pocket PAL - 1 digital refractometer (Attago, Ltd, Japan), with a scale ranging from 0 to 53% (°Brix) and automatic temperature compensation.

3.5.3 Total Titratable Acidity (ATT)

Titratable acidity was determined in triplicate using 5.0 g of the papaya sample and adding 50 mL of distilled water, which was then homogenised. The mixture was then placed in an Erlenmeyer flask, three drops of 1% alcoholic phenolphthalein were added and titrated with a previously standardised 0.1N NaOH solution until the turning point was reached, which was characterised by the appearance of a persistent pink colour. The results were expressed as a percentage of citric acid (IAL, 1985).

The formula used to calculate the acidity percentage was:

$$Acidez(\%) = \frac{F_{ácido} x\ 10\ x\ F_{NaOH} x\ V_{NaOH}}{M}$$

F_{acid} : Correction factor obtained by standardising citric acid

F_{NaOH} : Correction factor obtained from NaOH standardisation

V_{NaOH} : Volume spent titrating the sample (mL)

M: mass of pulp (g)

3.5.4 TSS/ATT ratio

The method is based on calculating the ratio of TSS (°Brix)/(ATT) Total titratable acidity, expressed as organic acid predominantly in the fruit, which measures the degree of sweetness of the fruit.

3.5.5 Statistical analysis

Analyses of variance were carried out using the Sanest statistical analysis programme (ZONTA; MACHADO, 1984).

CHAPTER 4

RESULTS AND DISCUSSION

4.1 . Total Soluble Solids

Table 9 shows the average TSS values (°Brix) of papaya fruit as a function of the vegetable ash and carnauba bagana treatments.

Tables 9 - Total soluble solids values (°Brix) of papaya fruit.

Grey (C)	Bagana (B)	Blocks			
		I	II	III	IV
C0	B0	10,87	10,74	9,80	10,51
	B1	9,27	10,57	10,30	10,70
C1	B0	10,38	10,53	10,40	10,23
	B1	9,80	10,40	10,53	11,77
C2	B0	9,97	10,71	10,73	10,75
	B1	10,70	9,93	11,10	11,00
C3	B0	10,31	10,32	10,47	10,20
	B1	10,40	10,13	10,80	10,63

Source: Author (2015).

The results for total soluble solids obtained in this research are within the range of reference values, as can be seen in the research carried out by Oliveira (1999), who obtained TSS values of around 11.5 per cent for the 'Improved Sunrise Solo Line 72/12' cultivar and 10.4 per cent for the 'Tainung 01/781' hybrid.

Souza (2004) obtained average TSS values for the Tainung 01 cultivar of around 11.1% at the second stage of ripeness, a result that is in line with those obtained in this research.

The average TSS values illustrated in Graph 1 show that for the treatments with ash application, the presence of mulch increases the total soluble solids values in papaya fruit. Similar results were obtained by Sousa (2015) when he analysed quantitative variables of the papaya tree in this same experimental unit.

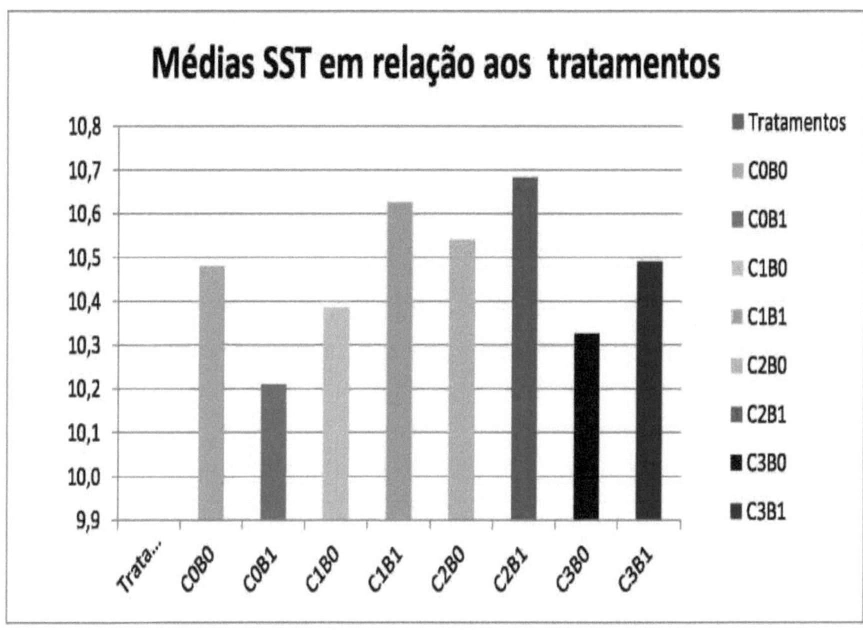

Graph 1 - Average TSS values according to treatments.

Source: Author (2015).

The analysis of variance for the TSS response variable (Table 10) indicated significant effects for the inputs vegetable ash and carnauba bagasse with Prob. > F values of more than 50 and 70%, respectively, therefore with very high probabilities of error in rejecting the null hypothesis.

Table 10 - Analysis of variance diagram for TSS

Cause of Variation	G.L	S. Q.	Q. M.	F	Prob>F
Blocks	3	1.0855341			

Grey	3	0.3254093	0.1084698	0.7938	0.52930
Waste (a)	9	1.2297525	0.1366392		
Instalments	15	2.6406959			
Bagana	1	0.0385032	0.0385032	0.1404	0.71453
CIN x BAG	3	0.3175596	0.1058532	0.3859	0.76745
Residue(B)	12	3.2916874	0.2743073		
Total	31	6.2884461			

Source: Author (2015).

4.2 Total Titratable Acidity

Table 11 shows the average total titratable acidity (TTA) values of papaya fruit as a function of the vegetable ash and carnauba bagana treatments.

Table 11 - Total titratable acidity values (%) of papaya fruit

Grey (C)	Bagana (B)	Blocks			
		I	II	III	IV
C0	B0	0,05	0,08	0,04	0,05
	B1	0,06	0,05	0,04	0,06
C1	B0	0,06	0,05	0,05	0,05
	B1	0,06	0,05	0,06	0,08
C2	B0	0,05	0,06	0,05	0,05
	B1	0,06	0,05	0,06	0,05
C3	B0	0,05	0,05	0,06	0,06
	B1	0,05	0,05	0,06	0,05

Source: Author (2015).

The average total titratable acidity values of the fruit analysed ranged from 0.04 to 0.08%, making it a very low acidity. There were also low variations between treatments, although these values were below the reference values. Souza (2004), in a study of formosa papaya, found ATT values of between 0.16 and 0.17% in the second stage of ripeness. The response variable acidity, according to the analysis of variance (Table 12), showed no statistically significant difference between the treatments for the standard levels used in conventional statistics.

Table 12- Analysis of variance for total titratable acidity

Cause of Variation	G.L.	S. Q.	Q. M.	F	Prob>F
Blocks	3	0.0000594			
Grey	3	0.0000844	0.0000281	0.2523	0.85809
Waste (a)	9	0.0010031	0.0001115		
Instalments	15	0.0011469			
Bagana	1	0.0000281	0.0000281	0.3333	0.58005
CIN x BAG	3	0.0002094	0.0000698	0.8272	0.50603
Residue(B)	12	0.0010125	0.0000844		
Total	31	0.0023969			

Source: Author (2015).

4.3 TSS/TPA ratio

Table 13 shows the average values for the TSS/ATT ratio of papaya fruit as a function of the vegetable ash and carnauba bagana treatments.

The data shows that the average values for the TSS/ATT ratio varied between 173.49 and 201.87 with the treatments applied. However, these values are well above some studies with Formosa Group papaya, such as that carried out by Rodolfo Júnior et al. (2007), who obtained average values of around 100.10.

Table 13 - TSS/ATT ratio values

Grey (C)	Bagana (B)	Blocks			
		I	II	III	IV

C0	B0	217,4	134,4	245,0	210,2
	B1	154,5	211,4	257,5	178,3
C1	B0	173,0	210,6	208,0	204,6
	B1	163,3	208,0	175,5	147,1
C2	B0	199,4	178,5	214,6	215,0
	B1	178,3	198,6	185,0	220,0
C3	B0	206,2	206,4	174,5	170,0
	B1	208,0	202,6	180,0	212,6

Source: Author (2015)

The analysis of variance indicated a non-significant effect at the 5% probability level for the TSS/TTA ratio at the different levels of treatment with vegetable ash and bagana, indicating that the application of these inputs was not a determining factor in the significant variation in the values of this response variable (Table 14).

Table 14 - Analysis of variance diagram for the TSS/ATT ratio

Cause of Variation	G.L	S. Q.	Q. M.	F	Prob>F
Blocks	3	1259.5755731			
Grey	3	1010.7397658	336.9132553	0.2986	0.82659
Waste (a)	9	10154.4367512	1128.2707501		
Instalments	15	12424.7520902			
Bagana	1	235.6076224	235.6076224	0.3570	0.56706
CIN x BAG	3	1421.6760697	473.8920232	0.7182	0.56242
Residue(B)	12	7918.5177710	659.8764809		
Total	31	22000.5535534			

Source: Author (2015).

4.4 . pH analysis

Table 15 shows the average pH values of papaya fruit as a function of the vegetable ash and carnauba bagana treatments.

Table 15 - Averages of pH values of papaya fruit as a function of treatment.

Grey (C)	Bagana (B)	pH			
		I	II	III	IV
C0	B0	5,34	5,52	5,57	5,43
	B1	5,58	5,69	5,43	5,62
C1	B0	5,41	5,27	5,43	5,55
	B1	5,55	5,43	5,87	5,25
C2	B0	5,6	5,4	5,5	5,5
	B1	5,56	5,51	5,55	5,59
C3	B0	5,43	5,45	5,19	5,67
	B1	5,58	5,46	5,61	5,25

Source: Author (2015).

The pH values obtained in this study considering the average values for all treatments (primary and secondary) is 5.5; therefore, similar to those obtained by

Rodolfo Júnior et al. (2007), who found pH values of 5.2 for the Formosa cultivar and 5.4 for the S. Solo cultivar. The results show that the pH values of papaya fruit are within the recommended range for fresh consumption.

The analysis of variance (Table 16) indicated a non-significant effect at the 5% probability level for the pH variable at the different levels of treatment with vegetable ash and the use of carnauba bagasse.

Table 16- Analysis of variance diagram for the pH variable

Cause of Variation	G.L.	S. Q.	Q. M.	F	Prob>F
Blocks	3	0.0133094			
Grey	3	0.0315844	0.0105281	0.6745	0.59142
Waste (a)	9	0.1404779	0.0156087		
Instalments	15	0.1853717			
Bagana	1	0.0504032	0.0504032	1.6087	0.22722
CIN x BAG	3	0.0089593	0.0029864	0.0953	0.96052
Residue(B)	12	0.3759877	0.0313323		
Total	31	0.6207220			

Source: Author (2015).

As highlighted above, Sousa (2015), analysing variables associated with physical productivity and water productivity in this same experimental unit, obtained highly significant results for the carnauba bagana production factor. At first, it is surprising that there are no significant probability levels for rejecting the null hypothesis when analysing qualitative variables in papaya fruit, as observed in this study.

However, these qualitative variables are influenced by factors such as soil and climate conditions, variety, time and place of harvest, cultivation and post-harvest handling, according to Fagundes and Yamanishi (2001), cited by Rodolfo Júnior et al. (2007).

These results may have been due to the fact that the plant was in a system that didn't suffer any stresses or because these variables are a genetic characteristic of the cultivar itself, where the genes that control this function are not easily altered by climatic factors, only if there had been a great deal of stress, such as water stress, mechanical damage or a lack of nutrients, where the plant wouldn't have been able to complete its biological cycle.It is also important to emphasise that the main objective of plants is the perpetuation of their species, so the plant will try its best to produce quality seeds and consequently a quality fruit because the fruit is the place where the seeds are protected, which is directly in contact with the mesocarp (pulp) and the endocarp, which is a large part of the fruit.

CHAPTER 5

CONCLUSION

1. The fruit of the Formosa Group showed average values associated with the qualitative variables total soluble solids (TSS), total titratable acidity (TTA), pH and TSS/TTA ratio, which give them qualities within the standards for commercialisation on the domestic market. In addition, non-formal information obtained from consumers of this product is very positive about the quality of the fruit, which has helped the farmer to sell his product at an attractive price (R$2.00 $kg^{-1)}$.
2. The degree of sweetness of papaya fruit measured by the TSS/ATT ratio gives average values, if all the treatments (primary and secondary) are taken into account, that are up to double the average reference value for Formosa Group fruit.
3. The non-significance of the primary and secondary treatments related to the qualitative variables of the papaya fruit analysed in this research is highly significant from the point of view of the allocation of financial resources in the factor-product relationship.
4. The analyses related to the application of the plant ash production factor cannot be considered absolutely decisive due to problems associated with the application of this input during the experiment.
5. A balance of nutrients required and made available, especially nitrogen, points to the need to adjust the production system in order to maximise other qualitative variables of interest.

CHAPTER 6

REFERENCE

AQUINO, A. M de; ASSIS, RL de. Agroecology: principles and techniques for sustainable organic agriculture. **Brasília: Embrapa Informação Tecnológica**, 2005.

ASSIS, R. L de.; ROMEIRO, A. R. Agroecology and organic agriculture: controversies and trends. **Desenvolvimento e meio ambiente**, Editora UFPR, v. 6, p. 67-80, 2002.

AYERS, R. S.; WESTCOT. D. W. **Water quality in agriculture**. Translation by GHEYI, H. J.; MEDEIROS, J. F. de; DAMASCENO, F. A. V. Campina Grande: UFPB, 1999 (FAO Studies: Irrigation and Drainage, 29 Revised I). Original title: Water Quality for Agriculture.

BRAZIL. Law n°11.326, of 24 June 2006, establishes the guidelines for the formulation of the National Policy for Family Farming and Rural Family Enterprises. **Diário Oficial da União**, Poder Legislativo, Brasília, DF, 25 June 2006. Section 1, p 1.

BUAINAIN, A. M. Family farming, agroecology and sustainable development: Issues for debate. nov/06. **Inter-American Institute for Co-operation on Agriculture**. -Brasília: IICA ,v.5 , 2006.

BUAINAIN, A. M. et al. Family farming: a study of regional targeting. In: **Congress of the Brazilian Society of Rural Economics and Sociology**. 2004.

CABRAL, R. C. **Reference evapotranspiration from Hargreaves (1974) corrected by the Penman-Monteith/FAO method (1991) for the state of Ceará.** 2000. 83 f. Dissertation (Master's Degree in Irrigation and Drainage) - Department of Agricultural Engineering, Federal University of Ceará. Fortaleza, CE, 2000.

CAMPANHARO, M. et al. **Use of wood ash as a soil corrective .**

FertBio:Challenges for efficient land use and environmental quality,2008.

CAPORAL, F. R.; COSTABEBER, J. A. Agroecology: some concepts and principles. 24 p. Brasília: MDA. SAF/DATER-IICA, 2004.

CARVALHO,R. I. N.; FIORAVANÇO, J. C.; PAIVA, M. C.; MANICA, I.**Physical and chemical characteristics of papaya commercialised in Porto Alegre-RS.** Rev. Brás. Frutic. 14(1): 143-147, 1992.

CHITARRA, M. I. F.; CHITARRA, A. B. **Post-harvest of fruit and vegetables: physiology and management**. 2. ed. Lavras, 2005. 785p.

CRISÓSTOMO, L. A.; NAUMOV,A. Fertilising for high productivity and quality: tropical fruit trees in Brazil. - Fortaleza : Embrapa Agroindústria Tropical, 2009. 238 p.; 21 cm. - (IIP. Bulletin 18).

DALMOLIN, R. S. D.; GONÇALVES, C. N.; KLAMT, E.; DICK, D. P. Relationship between soil constituents and their spectral behaviour. **Ciência Rural**, Santa Maria, v. 35, n. 2, p. 481 - 489, 2005.

DANTAS, J. L. L.; CASTRO NETO, M. T. **Papaya production:** technical aspects. Cruz das Almas: Embrapa Cassava and Fruit Growing; 2000. 77p.

EHLERS, E. Agricultura sustentável: origens e perspectivas de um novo paradigma - São Paulo: Livros da Terra 1996.

EMBRAPA. National Research Centre for Tropical Agroindustry (Fortaleza, CE), **Data climatological**: Pentecoste Station, 1998. Fortaleza:Embrapa-CNPAT/UFC, Embrapa: Agrometeorological Bulletin No. 16. p. 1 - 14, 1999.

EMBRAPA, Communication for Technology Transfer. Embrapa Cassava and Fruit Growing (Cruz das Almas, BA). **Papaya. Production: technical aspects** / Aldo Vilar Trindade, organiser - Brasília, 2000.

FAO - Food and Agriculture Organisation of the United Nations. Statistical databases agriculture. 2010. Available at: < http://faostat.fao.org/

>.Accessed 2015.

FAGUNDES, G. R.; YAMANISHI, O. K. Physical and chemical characteristics of 'solo' group papaya fruit marketed in 4 establishments in Brasília-DF. **Revista brasileira de fruticultura**, v. 23, n. 3, p. 541-545, 2001.

FERREIRA, J. P. **Experimental precision for papaya cultivation in the field.** 2014.Dissertation (Master's Degree in Tropical Agriculture) - Centro Universitário Norte do Espírito Santo, Universidade Federal do Espírito Santo, São Mateus,2014.

IBGE - Brazilian Institute of Geography and Statistics. **Agricultural census 2006**. <LINK:http://cod.ibge.gov.br/1XU4R>, accessed in 2015.

GOMES, J. A. F. **Evaluation of the potential of carnauba bagana for feeding sheep in north-eastern Brazil**. 2008. 71 f. Dissertation (Master's in Zootechnics) - Centre for Agrarian and Biological Sciences, Vale do Acaraú State University, Sobral, 2008.

GLIESSMAN, S. R. **Ecological processes in sustainable agriculture.** Porto Alegre: Federal University of Rio Grande do Sul, 2000. 637 p.

GLIESSMAN, S. R. **Agroecology: ecological processes in sustainable agriculture.** 2. ed. Porto Alegre: Ed. da UFRGS, 2001. 653 p.

KAMIYAMA, A. Caderno Educação Ambiental - Agricultura sustentável. V.13. Secretariat for the Environment / Coordination of Biodiversity and Natural Resources. São Paulo: SMA, 2011.

MARINHO, C. S. et al. Nitrogen sources and doses and the quality of papaya fruit. **Scientia Agricola**, v. 58, n. 2, p. 345-348, 2001.

MONTENEGRO, A. A. T.; BEZERRA, F. M. L.; LIMA, R. N. Evapotranspiration and crop coefficients of papaya for the coastal region of Ceará. **Engenharia Agrícola**, Jaboticabal, v. 24, n° 2, p. 464 - 472, 2004.

OLIVEIRA, A. M. G. et al. Nutrition, liming and fertilisation of irrigated papaya. **Cruz das Almas: Embrapa Cassava and Fruit Growing**, 2004.

OLIVEIRA, A. M. G.; SOUZA, L. F. da S.; COELHO, E. F. Mamoeiro. In: CRISOSTOMO, L. A.; NAUMOV, A. (Org.). (**Fertilising for high productivity and quality: tropical fruit trees in Brazil**. 1ed. Fortaleza: Embrapa Agroindustria Tropical; Instituto da Potassa, 2009, v., p. 146-165.

OLIVEIRA, F. N. S. et al. Influence of mulch on the development of tropical fruit trees - Fortaleza : **Embrapa Agroindústria Tropical**, (Documentos, 49), 2002.

PADOAN, L. L. F. . Analysis of alternative agricultural practices: Case studies of the Vicente Nica Alternative Agriculture Centre and the Peixe Cru Settlement. In: II Convibra: Agronomy, 2014. Online Proceedings, 2014.

PAULUS, G. **From the Modern Standard to Alternative Agriculture: Possibilities for Transition**. 1999. Dissertation (Master's Degree in Agroecosystems) - Agricultural Sciences Centre, Federal University of Santa Catarina, Florianópolis, 1999.

PIVA, R . et al. Maintenance fertilisation of Bordô vines using vegetable ash and bovine manure in an organic system. **Rev. Bras. Ciênc. Agrár. Recife**, v. 9, n. 2, p. 219-224, 2014.

QUEIROZ ,R. F. **Desenvolvimento de Mamão Formosa 'Tainung 01' Cultivado em Russas - Ceará**.2009.Dissertação (Mestrado em Agronomia:Fitotecnia) - Universidade Federal Rural do Semi-Árido, Mossoró- RN ,2009.

RODOLFO JÚNIOR, F. et al. Physical and Chemical Characterisation of Papaya Fruits Sold at Empasa in Campina Grande-PB. **Revista Brasileira de Produtos Agroindustriais**, v. 9, n. 1, p. 53-58, 2007.

SANCHES, J. (2003) **Post-harvest of papaya**, In: Informe-on-line Toda Fruta. Edition of 24/09/2003, available at: <http:// www.todafruta.com.br>.

SERRANO, L. A. L.; CATTANEO.L. F. O Cultivo do Mamoeiro no Brasil.**Revista Brasileira de Fruticultura,** v. 32, n. 3 p. 657-959.

SOUSA, P. G. S. **Mamoeiro productivity in response to vegetable ash and carnauba bagana inputs in the context of agroecological family farming**. 2015. Monograph (Graduation in Agronomy)-Centre of Agrarian Sciences, Federal University of Ceará, Fortaleza, 2015.

SHINAGAWA, F. B. **Evaluation of the biochemical characteristics of papaya pulp (Carica papaya L.) processed by high hydrostatic pressure**. 2009. Master's dissertation in Chemical and Biochemical Process Technology, School of Chemistry, UFRJ, RJ, 2009.

SOFIATTI, V. et al. Wood ash and sewage sludge as a source of nutrients for cotton growth. **Revista de Biologia e Ciências da Terra**, v. 7, n. 1, p. 144-152, 2007.

SOUZA, L. M. de. **Some physical and chemical characteristics of papayas (Carica papaya L.) from the "formosa" (Tainung 01) and "SOLO" groups.**

(Golden), with and without physiological spot, harvested at different stages of ripeness. Dissertation (Master's Degree) - Universidade Estadual do Norte Fluminense, Campos dos Goytacazes - RJ, 82p. 2004.

TRINDADE, A. V. et al. **Papaya. Production: technical aspects**. Embrapa Cassava and Fruit Growing (Cruz das Almas, BA). - Brasília: Embrapa Comunicação para Transferência de Tecnologia, 2000.

ZONTA, E.P.; MACHADO, A. A. Statistical analysis system for microcomputers - Sanest. Pelotas: UFPel, 1984. 109p.

Annex I (Photos throughout the experiment)

Preparing the soil
Source: Author (2015).

Plant growth
Source: Author (2015).

Flowering and beginning of production.
Source: Author (2015).

Papaya growth
Source: Author (2015).

Production
Source: Author (2015).

Production depending on the different treatments
Source: Author (2015).

Printed by Books on Demand GmbH, Norderstedt / Germany